Snowflake Follies

QUILTS TO MAKE IN A WINTER WEEKEND

Terry Martin

Martingale®
& COMPANY

CREDITS

President • NANCY J. MARTIN
CEO • DANIEL J. MARTIN
Publisher • JANE HAMADA
Editorial Director • MARY V. GREEN
Managing Editor • TINA COOK
Technical Editor • CYNDI HERSHEY
Copy Editor • MELISSA BRYAN
Design Director • STAN GREEN
Illustrator • LAUREL STRAND
Cover and Text Designer • TRINA STAHL
Photographer • BRENT KANE

That Patchwork Place® is an imprint of Martingale & Company®.

Snowflake Follies: Quilts to Make in a Winter Weekend
© 2003 by Terry Martin

Martingale & Company
20205 144th Avenue NE
Woodinville, WA 98072-8478 USA
www.martingale-pub.com

Printed in China
08 07 06 05 04 03   8 7 6 5 4 3 2 1

Library of Congress Cataloging-in-Publication Data

Martin, Terry.
    Snowflake follies / Terry Martin.
        p.   cm.
    ISBN 1-56477-480-5
1. Patchwork—Patterns.   2. Quilting.   3. Snowflakes in art.   I. Title.
    TT835.M38425   2003
    746.46'041—dc21

                                    2003004864

MISSION STATEMENT

Dedicated to providing quality products and service to inspire creativity.

Dedication

I WOULD LIKE to dedicate this book to my parents, Richard "Red" Maxwell and Alberta "Bert" Maxwell. I love you, Mom and Dad.

Acknowledgments

THIS BOOK WOULD not have been possible without the unflinching support of my husband, Ed, and my daughter, McKenzie. They have endured a number of nights when dinner consisted of grilled cheese sandwiches and soup, and I didn't even cook that, Ed did. Thank you both so much for letting me stretch my creative wings and fly under a cape of fabric!

I would also like to thank Barb Dau for her extraordinary talent as a long-arm machine quilter and artist. Barb, thank you for coming to my rescue and making my work sing.

I would like to thank all the folks at Martingale & Company for their support and work on this book; it's beautiful. This is the third book we have created together and your help and advice have been immeasurable!

Contents

Preface

What a wonderful time I have had working on these projects and writing this book! It's been fun and very rewarding to make an artistic study of the beautiful snowflake from the quilter's point of view. I loved the challenge of really stretching my artistic and quilting imagination while focusing on one design element.

Every time I think of snowflakes I smile and the memories flood in: from my childhood, catching snowflakes on my tongue and being terrified sledding down the steep hill next to our house on my uncontrollable sled, to watching my daughter, McKenzie, grow up playing in the snow. We stay up past her bedtime when it snows, even on a school night, and snuggle under a quilt to watch the snowfall highlighted by the streetlight in front of our house. McKenzie and I are always amazed at how quiet and serene the world becomes when it snows.

We don't get a lot of snow here in the Pacific Northwest, except in the Cascade and Olympic Mountains, but when we do get a few inches on the ground the entire region shuts down! You'd think that with our vast experience driving in the rain, we could adapt to a bit of snowfall; instead, general chaos ensues. Our poor weathermen try to predict our perfectly unpredictable weather, especially snowfall, so whatever they forecast, we assume the opposite. Even though our weather is such a constant surprise, the grocery stores are still raided in the attempt to stock up for the moment when the big snow is announced. It might only be a few inches and last a couple of days at the most, but we Northwesterners will be prepared.

When the schools are shut down, the kids head to the nearest hill for sledding. I love watching children and adults of all ages pile onto their sleds (sometimes even three deep!) and slide down the hill behind our house. Now snowboards are the "big thing" and there is always one ramp made by piled-up snow for the hotshots to fly up and over. They refuse to come in until their faces have turned blue! McKenzie's favorite warm-up treat is white hot chocolate with a peppermint stick for stirring. She knows how to make it for herself while I'm busy at the sewing machine.

I hope you enjoy reliving fond memories of snowfall while you are making one of these projects or while designing your own.

Terry Martin

Introduction

WHO DOESN'T LIKE a snowflake? Can you count how many times you as a child—or as an adult—have marveled at the beauty of new falling snow? Can you count the hundreds of snowflakes you cut out of paper as a child? I didn't think so. This book is full of projects that will awaken sweet memories of your childhood and give you the notion to capture them in a quilt that, unlike the delicate snowflake, will last forever.

Snowflakes are fabulous, intricate, lacey, strong, delicate, beautiful, exquisite, and a bunch of fun to play with from a design perspective! And after making more than a dozen projects with snowflakes as a design element, I have discovered that the simpler the snowflake and design, the prettier and more striking the quilt. I would never have thought I'd come to that conclusion when I first started playing with snowflakes, fabric, and thread. I assumed that the more complex and intricate the design, the better it would look; but it didn't work out that way!

That's when I discovered that I don't do complex and intricate well. I don't have the patience for it. And now I know why. Middle age. I find my attention span is now equal to my 15-year-old's, especially when I ask her to do chores. You know the blank glazed-over look they give you when they think you think they are paying attention to you. I would find myself in the middle of one project while designing the next project in my head. I couldn't wait to start the next project but, having some discipline, I told myself to finish at least the top of the quilt I was presently working on before starting something new. With this "rule" to my quilting, I simplified the design and assembly steps so that I could accomplish them quickly. I like the get-it-done-in-a-weekend concept, and the simplified designs were still working for me. Sticking to the things I know and love has resulted in this book.

Another thing I discovered is that using the traditional method of cutting out snowflakes with paper and scissors doesn't work very well with fabric and scissors. With the fabric being thicker than paper I couldn't get the perfect crease when folding, so the snowflakes weren't perfectly formed. Needless to say, I didn't use them in the projects in this book.

What I did discover about designing snowflakes for quilts is that using simple shapes and creating a segmented snowflake opened up all sorts of possibilities. I have included a section about creating snowflakes in this manner so that you can make projects with your own simple but elegant snowflake designs.

I am also simply amazed at all the fabulous snowflake fabric that's on the market these days. I have snowflake fabric in just about every color under the sun, including rainbow-colored hues. I have contacted several fabric manufacturers and they assure me they will continue to produce snowflake fabric. There is an abundance of subtle, cheerful, sophisticated, beautiful, and lively snowflake prints to choose from, so have fun!

From the first fall frost to the welcome spring thaw, I hope that you will make and use these quilts as a fun change of decor or to snuggle under during those chilly winter nights. Although these patterns using snowflake fabric and designs may be seasonal, they will last a lifetime in your heart as they conjure up memories of winters past.

Crystal Blue

FINISHED QUILT SIZE: 26½" x 26½"
FINISHED BLOCK SIZE: 16"

I love working with batik fabrics. The crisp fabric is so agreeable to work with and the colors are so rich. It has been fun collecting all the wonderful snowflake-printed batiks. Any of the projects in this book would be fabulous using batik fabrics, and batiks look great mixed in with other cotton fabrics as well.

This block looks complicated, but when you break it down into its individual components you will see that it's just lots of half-square triangles! Slow down and take your time matching and pinning your points when stitching, and you will have a great snowflake project finished in a weekend. The border treatment is also fun and easy to do and I think it adds that extra touch to the center design.

MATERIALS

Yardage is based on 42"-wide fabric.

- ⅞ yard of dark blue snowflake-print batik for block, outer border, and binding
- ½ yard of light blue snowflake-print batik for block and outer border
- ⅜ yard of gray snowflake print for block and inner border
- ⅜ yard of white snowflake print for block
- ⅞ yard of fabric for backing
- 30" x 30" piece of batting

CUTTING

All measurements include ¼"-wide seam allowances.

From the dark blue snowflake print, cut:

- ♦ 2 strips, 2⅞" x 42"; crosscut into 16 squares, 2⅞" x 2⅞". Cut each square once diagonally to yield 32 half-square triangles.
- ♦ 3 strips, 2½" x 42"; crosscut into 36 squares, 2½" x 2½"
- ♦ 1 strip, 4½" x 42"; crosscut into 4 rectangles, 4½" x 10½"
- ♦ 3 strips, 2½" x 42"

From the light blue snowflake print, cut:

- ♦ 2 strips, 2⅞" x 42"; crosscut into 16 squares, 2⅞" x 2⅞". Cut each square once diagonally to yield 32 half-square triangles.
- ♦ 3 strips, 2½" x 42"; crosscut into:
    - 16 rectangles, 2½" x 4½"
    - 4 squares, 2½" x 2½"

From the white snowflake print, cut:

- 1 strip, 2⅞" x 42"; crosscut into 8 squares, 2⅞" x 2⅞". Cut each square once diagonally to yield 16 half-square triangles.
- 1 strip, 2½" x 42"; crosscut into 16 squares, 2½" x 2½"
- 1 strip, 3⁵⁄₁₆" x 42"; crosscut into 4 squares, 3⁵⁄₁₆" x 3⁵⁄₁₆"

Tɪᴘ: *To measure a sixteenth of an inch, simply center the fabric edge between the ⅛" markings on your quilting ruler.*

From the gray snowflake print, cut:

- 1 strip, 2⅞" x 42"; crosscut into 4 squares, 2⅞" x 2⅞". Cut each square once diagonally to yield 8 half-square triangles.
- 1 square, 4½" x 4½"
- 2 strips, 1½" x 42"; crosscut into:
  - 2 rectangles, 1½" x 16½"
  - 2 rectangles, 1½" x 18½"

ASSEMBLY

1. For the center block and outer border, join half-square triangles as shown to make the following triangle squares:

- 20 from the dark blue and light blue batik triangles
- 8 from the dark blue batik and white print triangles
- 8 from the light blue batik and white print triangles

  Press the seams toward the darker fabric within each triangle-square unit.

Make 20.

Make 8.

Make 8.

2. Combine appropriate triangle squares from step 1 with four 2½" white squares as shown to make each of the corner units. Press the seams within each row in opposite directions. Join the rows and press the seams as shown.

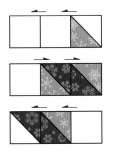 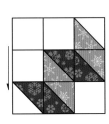

Make 4.

3. Sew the remaining half-square triangles of dark blue, light blue, and gray to the 3⁵⁄₁₆" white squares. Press the seams away from the center.

Make 4.

4. Combine the remaining triangle squares into pairs to form rectangular units. Press the seams in either direction. Sew each of these units to a unit from step 3. Press the seams toward the rectangular units.

 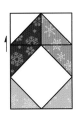

Make 4.

5. Add two units from step 4 to opposite sides of the gray print square, making sure the gray triangles are next to the gray center square. Press toward the center square.

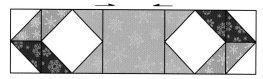

Make 1.

6. Join two of the corner units from step 2 to opposite sides of the remaining units from step 4. Press toward the corner units.

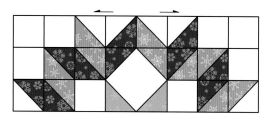

Make 2.

7. Add these large units to the remaining sides of the center square to complete the block. Press final seams in either direction.

BORDERS

1. Using the 1½" x 16½" strips of gray print, add the two side inner borders as described in "Quiltmaking Basics" on page 51. Repeat this step for the top and bottom borders using the 1½" x 18½" gray strips. Press the seams toward the borders.

2. Draw a pencil line diagonally on the back of 32 of the dark blue batik squares.

3. Pin one of these squares, wrong side up, on top of each end of a 2½" x 4½" light blue rectangle and stitch on the pencil line. Trim the seam allowance to ¼" and carefully press the seam toward the triangle. Note that you will need to make eight with the triangles facing one direction and eight with them facing the opposite direction.

Make 8.   Make 8.

4. Join these units into pairs.

Make 4.   Make 4.

5. Combine the remaining four dark blue and four light blue batik squares with the remaining eight triangle squares to make corner units for the outer border. Press.

Make 4.

6. Join the units from step 4 to each end of a dark blue batik rectangle, referring to the illustration and to the quilt photo on page 9 for proper color position. Press toward the rectangles.

Make 4.

7. Join two of the borders to the top and bottom of the quilt. Press toward the inner border.

8. Add the corner units created in step 5 to each end of the two remaining borders, taking care to position the corner units correctly. Press the seams toward the corners.

Make 2.

9. Join the borders to each side of the quilt top. Press toward the inner border.

FINISHING

1. Make a quilt sandwich with the quilt top, batting, and backing; for more specifics on finishing techniques, refer to "Quiltmaking Basics" on page 51. Baste.

2. Quilt as desired. My machine quilting is still at the beginning level, so I stitch in the ditch to anchor the layers together and then I highlight the design of the block and quilt.

3. Trim the batting and backing even with the edges of the quilt top. Join the 2½" x 42" dark blue strips end to end for the binding; sew the binding to the quilt.

4. Add a hanging sleeve, if desired, and label your quilt.

Snow on the Cabin

Finished Quilt Size: 78½" x 90½"
Finished Block Size: 12"

You can find a mountain of snowflake-print fabric available in quilt shops these days. Even though it is considered a seasonal fabric, many shops maintain a small section devoted to holiday fabrics year-round, and I get a kick when I discover a new snowflake print! It was my lucky day when I found the teal and red fabrics used in this project. Most snowflake fabrics are white or blue, so I considered this a real find.

This quilt goes together very quickly. I like the way the blocks dance across the quilt at an angle, adding movement and a secondary pattern. This variation of the Log Cabin block is especially fun and easy to put together. My first experience with the traditional Log Cabin block was not the best. I had to really concentrate on which side the next log went on and believe me, I had to keep my seam ripper handy! This variation of the Log Cabin block is foolproof because you add the logs on only two sides of the square, not all the way around like the traditional block. You might consider chain-piecing each layer, but I don't consider this any faster since you will have to trim the block as you add each log.

The borders go on just as quickly as the blocks go together. You will make a strata for the last three borders, an extremely simple step yet one that adds an additional design element. Stratas are strips of fabric joined lengthwise, and the stratas are then subcut into whatever units you need. Making stratas is quicker and, I find, more accurate than sewing individual pieces together.

MATERIALS

Yardage is based on 42"-wide fabric.

- 4 yards of teal snowflake print for blocks, borders, and binding
- 2 yards of red snowflake print for blocks and border
- 1⅛ yards of gold snowflake print for blocks and border
- 1⅛ yards of gray snowflake print for blocks
- ⅝ yard of cream snowflake print for blocks
- 5½ yards of fabric for backing
- 82" x 94" piece of batting

CUTTING

All measurements include ¼"-wide seam allowances.

From the gold snowflake print, cut:
- 9 strips, 2½" x 42"; crosscut into:
    30 rectangles, 2½" x 4½"
    30 rectangles, 2½" x 6½"
- 7 strips, 2" x 42"

From the cream snowflake print, cut:
- 4 strips, 4½" x 42"; crosscut into 30 squares, 4½" x 4½"

From the gray snowflake print, cut:
- 13 strips, 2½" x 42"; crosscut into:
    30 rectangles, 2½" x 6½"
    30 rectangles, 2½" x 8½"

TIP: *Change the size of the quilt! If you are pulling from your stash, simply make as many 12" Log Cabin blocks as your fabric will allow. Play with the arrangement of the blocks to create a different look for your quilt. Add borders to frame the blocks, or leave them off; these blocks can stand on their own without borders!*

From the red snowflake print, cut:
♦ 18 strips, 2½" x 42"; crosscut into:
   30 rectangles, 2½" x 8½"
   30 rectangles, 2½" x 10½"
♦ 8 strips, 2" x 42"

From the teal snowflake print, cut:
♦ 20 strips, 2½" x 42"; crosscut into:
   30 rectangles, 2½" x 10½"
   30 rectangles, 2½" x 12½"
♦ 16 strips, 3½" x 42"
♦ 9 strips, 2½" x 42"

ASSEMBLY

1. Sew a 2½" x 4½" gold snowflake-print rectangle to the top edge of each cream snowflake-print square. Press toward the rectangle.

Make 30.

2. Sew a 2½" x 6½" gold snowflake-print rectangle to the left side of the unit created in step 1. Press toward the rectangle.

Make 30.

3. Continue adding the gray, red, and teal snowflake-print rectangles in order of increasing size until the block is complete. Press toward the new rectangle after each addition.

Make 30.

4. Sew the blocks into six rows of five blocks each. Press the seams in alternating directions from row to row.

5. Join the rows. Press final seams in either direction.

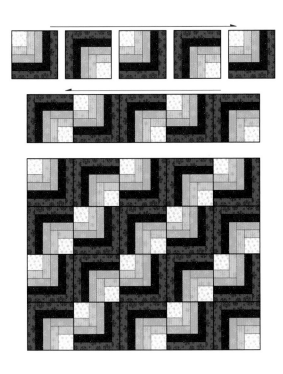

BORDERS

1. Sew the 2" x 42" gold snowflake-print strips end to end. Following the directions in "Quiltmaking Basics" on page 51, cut and sew the side borders to the quilt top. Repeat for the top and bottom borders. Press the seams toward the borders.

2. Join the 2" x 42" red snowflake-print strips end to end. Press the seams in either direction.

3. Join eight of the 3½" x 42" teal snowflake-print strips end to end. Repeat with the remaining eight strips. Press the seams in either direction.

4. Create a strip set (strata) by joining the teal snowflake-print strips to each side of the red snowflake-print strip. Press toward the teal strips.

5. Cut and sew the side borders from the strip set, following the directions in "Quiltmaking Basics" on page 51. Repeat for the top and bottom borders. Press the seams toward the borders.

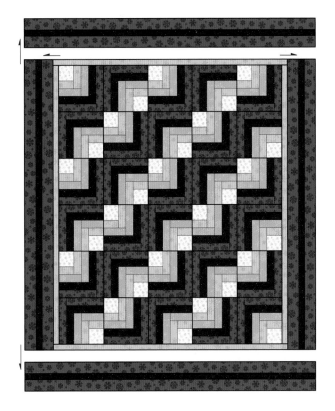

FINISHING

1. Make a quilt sandwich with the quilt top, batting, and backing as shown in "Quiltmaking Basics" on page 51. Baste.

2. Quilt as desired. My friend and long-arm machine quilter, Barb Dau, created a wonderful allover quilting pattern for this quilt. It looks like a swirling snowstorm.

3. Trim the batting and backing even with the edges of the quilt top. Join the 2½" x 42" teal strips end to end for the binding; sew the binding to the quilt.

4. Add a hanging sleeve, if desired, and label your quilt.

Puttin' on the Ritz

FINISHED QUILT SIZE: 29½" x 36½"
FINISHED BLOCK SIZE: 7"

I really like this little quilt and it was one of the first projects I designed for this book. I hope you like it as much as I do. I can just see the little snowflakes all dressed up with their tiny top hats and tails dancing like Fred Astaire as they fall gently to the ground. (Whew! Can I use my imagination or what?!)

The black-and-gold fabric has a sophisticated air that would look great in a den or an office, and I think this quilt would be a wonderful gift for a guy. To make this a very quick project, and since I didn't edgestitch the designs, I used Steam-A-Seam 2 to fuse the snowflakes. Steam-A-Seam 2 is my favorite double-backed fusible web product. It's easy to trace on and it presses down easily and thoroughly with no lifting. Another thing I like is that after you've pressed Steam-A-Seam 2 onto fabric, cut out your motif, and peeled off the second paper, it has a tacky feel like the back of a sticky note, allowing you to reposition the motif on the background fabric as many times as you want before giving the project a final pressing.

You might want to machine stitch the edges of the snowflakes with a blanket or blind stitch, or use your favorite method of hand appliqué for a quilt that is more durable.

MATERIALS

Yardage is based on 42"-wide fabric.

- 2½ yards of black-and-gold snowflake print for snowflakes, block background, inner and outer borders, backing, and binding
- 1 yard of tan-and-gold snowflake print for snowflakes, block background, and middle border
- ⅜ yard of Steam-A-Seam 2 or other fusible web
- 33" x 40" piece of batting

CUTTING

All measurements include ¼"-wide seam allowances.

From the black-and-gold snowflake print, cut:

- 2 strips, 8" x 42"; crosscut into 6 squares, 8" x 8". Use remaining fabric to cut 6 of the snowflake design.
- 8 strips, 2" x 42"; crosscut into:
    2 rectangles, 2" x 24½"
    2 rectangles, 2" x 28½"
    2 rectangles, 2" x 29½"
    2 rectangles, 2" x 33½"
- 1 rectangle, 33" x 40"
- 4 strips, 2½" x 42"

From the tan-and-gold snowflake print, cut:

- 2 strips, 8" x 42"; crosscut into 6 squares, 8" x 8". Use remaining fabric to cut 6 of the snowflake design.
- 4 strips, 1½" x 42"; crosscut into:
    2 rectangles, 1½" x 26½"
    2 rectangles, 1½" x 31½"

2. Following the manufacturer's instructions for the fusible web, peel off the backing and fuse six snowflake designs to the wrong side of each fabric.

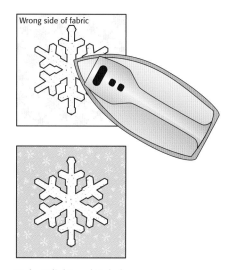

Make 6 light and 6 dark.

TIP: *For a very cute and scrappy doll quilt, use several 1930s-era reproduction fabrics and blanket stitch the snowflakes with a dark brown thread.*

TIP: *Fusible web is the greatest for quick appliqué, but sticky residue may collect on your iron if the web is even slightly exposed underneath the piece of appliqué when you press it in place. If your iron does not have a non-stick coating, try using a pressing cloth to protect against residue. If your iron does have a non-stick coating, make sure you clean the iron face before starting a new project!*

ASSEMBLY

1. Photocopy or trace the snowflake design on page 63. Use this as a template to trace a total of 12 snowflakes onto fusible web. This particular design is symmetrical so you will not need to make a mirror image of the design.

3. Cut out the snowflakes, remove the paper from the fusible web, and center each snowflake on the square of fabric that is the opposite of the fabric the snowflake is cut from. Once each snowflake is positioned to your liking, press with an iron per the directions of your fusible web.

Trace 12.

4. You may opt to stitch each of the snowflakes to the background using your favorite method.

Make 6.

Make 6.

5. Trim the blocks to 7½" x 7½", keeping the design centered.

6. Join the blocks in four rows of three alternating black and tan backgrounds. Press the seams toward the black backgrounds. Then join the rows to complete the quilt center. Press final seams in either direction.

BORDERS

1. Following the instructions in "Quiltmaking Basics" on page 51, add the inner border. Use the 2" x 24½" rectangles of black snowflake print for the top and bottom and the 2" x 28½" rectangles for the sides of the inner border.

2. Repeat step 1 for the tan middle border and the black outer border as shown.

FINISHING

1. Make a quilt sandwich with the quilt top, batting, and backing as shown in "Quiltmaking Basics" on page 51. Baste.

2. Quilt as desired. I carefully outline quilted each of the snowflakes with a gold metallic thread. If you use a metallic thread for machine quilting, choose the appropriate needle to help the stitching go smoothly. I use a topstitch needle or a metallic needle; the eyes are bigger, so the thread doesn't fray as it travels back and forth through the needle.

3. Join the 2½" x 42" black-and-gold strips end to end to make the binding; sew the binding to the quilt.

4. Add a hanging sleeve, if desired, and label your quilt.

Snowflakes in Redwork

Finished Quilt Size: 12¾" x 40½"
Finished Block Size: 6¼"

I love to sew, and you can usually find me in front of the sewing machine or at the ironing board. Working on this project was such a joy! It made me sit down and relax, and with needle, floss, and fabric, I rediscovered the true joy of handwork. Listening to music or the television, I was able to curl up in a corner of the couch and blissfully stitch away. I can hardly think of a better way to pass the time on a blustery winter day.

I happened to mention that I was working on a redwork design for this book, and Martingale's editorial director, Mary Green, suggested using a variegated thread. What a wonderful idea! As a cross-stitcher for twenty-plus years, I own every DMC floss made but I was never quite sure what to do with the variegated thread; this was an excellent way to experiment. I think the thread adds dimension to the design and I hope you agree.

MATERIALS

Yardage is based on 42"-wide fabric.

- ⅝ yard of red snowflake print for sashing, borders, and binding
- ⅜ yard of preshrunk white or cream muslin for blocks
- ⅜ yard of cream print for sashing and borders
- ½ yard of fabric for backing
- 17" x 45" piece of batting
- 2 skeins of DMC floss #75
- 5" to 6" embroidery hoop
- Embroidery needle

TIP: *As I mention in "Quiltmaking Basics," I don't generally wash my fabrics before cutting and piecing. However, I do recommend preshrinking the muslin for this project. Depending on thread count, muslin can shrink at a higher rate than other 100% cotton fabric.*

CUTTING

All measurements include ¼"-wide seam allowances.

From the muslin, cut:
- 1 strip, 7" x 42"; crosscut into 4 squares, 7" x 7"

From the red snowflake print, cut:
- 7 strips, 1½" x 42"
- 3 strips, 2½" x 42"

From the cream print, cut:
- 5 strips, 1½" x 42"

ASSEMBLY

1. Photocopy or trace each redwork snowflake pattern on pages 61–62.

TIP: *Photocopy machines can distort an image, so for consistent designs use the same machine for all your photocopying.*

2. Find the center of the muslin squares by folding each square in half in both directions and finger-pressing. Tape the paper designs to a tabletop. Center a muslin square over each design and tape the edges to anchor the muslin. Use your favorite method to trace the outside and inside of each snowflake design onto the muslin.

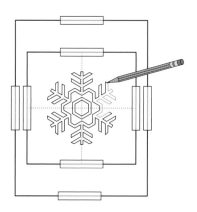

3. Center the marked muslin square in the embroidery hoop and tighten. Use two strands of DMC #75, each about 18" long, to embroider with a stem stitch or straight stitch directly over the marked lines of the design.

TIP: *Because you are using a variegated thread, the look of the embroidery will be different depending on which end of the floss you start with. With each 18" strand, you might be starting your stitching with a lighter or darker shade of the floss. Examine your cut pieces of floss and choose which end you want to start with. For example, your stitching with the first strand might have ended with a dark portion of the floss. If you begin the next strand with a much lighter portion, it will form a stark contrast. You'll probably want to begin using the strand from the darker end so that the joint isn't as noticeable.*

Reposition the hoop as needed. Don't stitch too close to the edge of the hoop because it could distort the design and, likewise, your stitches.

4. Trim the blocks to 6¾", centering the design.

5. Make three strip sets by joining a 1½" x 42" red print strip to each side of three of the cream print strips. These are A strip sets. Press the seams toward the red.

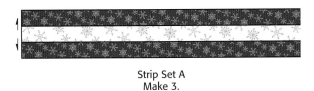

Strip Set A
Make 3.

6. Cut 13 segments, 6¾" wide, from two of the A sets.

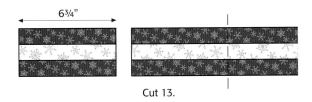

6¾"

Cut 13.

7. Cut 10 segments, 1½" wide, from the remaining A set.

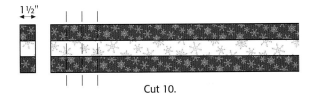

1½"

Cut 10.

8. Sew the 6¾" segments to the top of each redwork block. Also sew a segment to the bottom of just one redwork block. Press the seams toward the redwork segments.

Make 3.

Make 1.

9. Sew the blocks together into a long panel beginning and ending with a section of strip set A.

10. Make strip set B by joining a cream print strip to each side of the remaining 1½" x 42" red print strip. Press the seams toward the red.

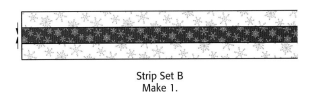

Strip Set B
Make 1.

11. Cut 20 segments, 1½" wide, from strip set B.

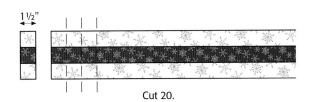

1½"

Cut 20.

12. Alternately join two 1½" segments of strip set B with one 1½" segment of strip set A to create nine-patch units. Press both seams toward the middle.

Make 10.

13. Sew a nine-patch unit to one end of each remaining 6¾" A segment. Press toward the A segment.

Make 8.

14. Sew four of these units together and add a remaining nine-patch unit to the end for a side border. Press the seams toward the A segments. Repeat to make the second side border.

Make 2.

15. Sew a border unit to each side of the redwork panel. Press the seams toward the borders.

FINISHING

1. Make a quilt sandwich with the quilt top, batting, and backing as shown in "Quiltmaking Basics" on page 51. Baste.

2. Quilt as desired. I machine quilted around the inside and outside of the snowflakes to make them stand out, and I stitched in the ditch within the border and sashing seams.

3. Trim the batting and backing even with the edges of the quilt top. Join the 2½" x 42" red print strips end to end for the binding; sew the binding to the quilt.

4. Add a hanging sleeve, if desired, and label your quilt.

Snow in the Mountains

FINISHED QUILT SIZE: 60½" x 80½"
FINISHED BLOCK SIZES: 8" and 48"

This snuggly flannel quilt is one of the coziest quilts I have made. Guests and family reach for it whenever they feel a chill; it's also one of the quilts they most frequently ask me to make for them!

I love one-block quilts; they simply fascinate me. Take an intricate-looking quilt block that seems too complex to stitch up as a 12" block, blow it up to 48" square, and all of a sudden it's not quite so intimidating! Everyone can handle 6½" squares—or bigger—to create a really large block. The other advantage to making a really large center block in this quilt is that it stitches up very quickly.

I added the Mountain block on the top and bottom to create a rectangular quilt and to help frame the large Arrow Crown block in the center.

Use your imagination! I think this quilt would look fabulous in large-print children's fabrics, plus it's the perfect size for a single bed!

Tip: *Flannel can be a little tricky to work with as it is usually a looser weave than other quilt fabrics and can easily stretch. Also, because of the excessive shrinking flannel is the only fabric other than muslin that I always preshrink. Take care not to handle your cut pieces too often or they might stretch out of shape; also, use a walking foot when you piece, and lengthen your stitch a little.*

MATERIALS

Yardage is based on 40"-wide fabric.

- 3 yards of burgundy snowflake-print flannel for the block, sashing, border, and binding
- 1¼ yards of holly-print flannel for blocks
- 1⅛ yards of cream print flannel for center block
- 1 yard of light green print flannel for blocks
- ⅝ yard of dark green print flannel for blocks
- 4 yards of fabric for backing
- 64" x 84" piece of batting

CUTTING

All measurements include ¼"-wide seam allowances.

From the burgundy snowflake print, cut:

- 1 strip, 6⅞" x 42"; crosscut into 4 squares, 6⅞" x 6⅞". Cut each square once diagonally to yield 8 half-square triangles.
- 1 strip, 13¼" x 42"; crosscut into 3 squares, 13¼" x 13¼". Cut each square twice diagonally to yield 12 quarter-square triangles.
- 3 strips, 3½" x 42"
- 7 strips, 6½" x 42"
- 8 strips, 2½" x 42"

From the holly print, cut:

- 2 strips, 6⅞" x 42"; crosscut into 8 squares, 6⅞" x 6⅞". Cut each square once diagonally to yield 16 half-square triangles.
- 1 strip, 12½" x 42"; crosscut into:
   - 1 square, 12½" x 12½"
   - 4 squares, 6½" x 6½"
- 1 strip, 9¼" x 42"; crosscut into 3 squares, 9¼" x 9¼". Cut each square twice diagonally to yield 12 quarter-square triangles.

From the cream print, cut:

- 2 strips, 6⅞" x 42"; crosscut into 8 squares, 6⅞" x 6⅞". Cut each square once diagonally to yield 16 half-square triangles.
- 1 strip, 6½" x 42"; crosscut into 4 squares, 6½" x 6½"
- 1 strip, 13¼" x 42"; crosscut into 2 squares, 13¼" x 13¼". Cut each square twice diagonally to yield 8 quarter-square triangles.

From the light green print, cut:

- 2 strips, 6½" x 42"; crosscut into 12 squares, 6½" x 6½"
- 1 strip, 9¼" x 42"; crosscut into 3 squares, 9¼" x 9¼". Cut each square twice diagonally to yield 12 quarter-square triangles.

From the dark green print, cut:

- 2 strips, 8⅞" x 42"; crosscut into 6 squares, 8⅞" x 8⅞". Cut each square once diagonally to yield 12 half-square triangles.

ASSEMBLY

1. Sew the eight burgundy print half-square triangles to eight holly-print half-square triangles.

2. Sew the eight remaining holly-print half-square triangles to eight cream print half-square triangles.

Make 8.   Make 8.

3. Combine the half-square-triangle units with the 6½" squares of holly print, cream print, and light green to form rows of three squares each. Press the seams toward the plain squares. Sew the rows together and press the seams away from the center row.

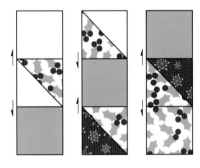

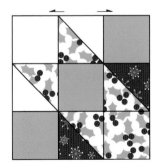

Make 4.

4. Sew eight burgundy print quarter-square triangles together with the cream print quarter-square triangles. Press the seams toward the burgundy triangles. Sew these units together to form 4 quarter-square-triangle units. Press the seams in either direction.

Make 4.

5. Sew a cream print half-square triangle to each short side of the four remaining burgundy quarter-square triangles. Press the seams toward the cream.

Make 4.

6. Sew a rectangle unit from step 5 to a cream print side of a quarter-square-triangle unit. Press the seams toward the square unit. Make four.

Make 4.

7. Sew two units from step 6 to opposite sides of the 12½" holly square, making sure the burgundy triangles are adjacent to the holly square. Press the seams toward the holly square.

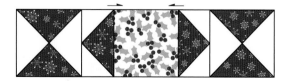

8. Sew the units from step 3 to opposite sides of the remaining units from step 6, referring to the illustration below for proper positioning. Press the seams away from the center.

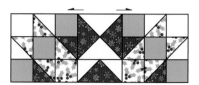

Make 2.

9. Sew the three sections together to complete the large center block. Press the seams in either direction.

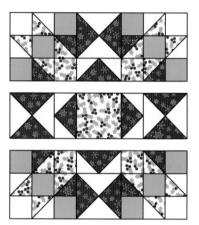

BORDERS

1. Sew the 3½" x 42" burgundy print strips together end to end.

2. Measure, cut, and sew the inner top and bottom borders to the center block as described in "Quiltmaking Basics" on page 51. Press the seams toward the border.

3. Sew the light green print and holly print quarter-square triangles together along short edges. Press the seams toward the light green.

Make 12.

4. Sew these units to the dark green print half-square triangles to complete the border blocks. Press toward the dark green.

Make 12.

5. Sew six of the border blocks together to form a pieced border. Repeat. Press the seams in either direction.

Make 2.

6. Sew a pieced border to the top and bottom of the center block, referring to the following illustration for proper positioning. Press the seams toward the inner border.

7. Sew the 6½" x 42" burgundy print strips together end to end.

8. Measure, cut, and sew the outer side borders to the quilt top as described in "Quiltmaking Basics" on page 51. Repeat for the outer top and bottom borders. Press the seams toward the outer borders.

FINISHING

1. Make a quilt sandwich with the quilt top, batting, and backing as shown in "Quiltmaking Basics" on page 51.

2. Quilt as desired. Quilter Barb Dau had a lot of fun with this project! She saw the 6" open blocks as a canvas to play with and essentially created a secondary design with her machine quilting.

3. Trim the batting and backing even with the edges of the quilt top. Join the 2½" x 42" burgundy print strips end to end for the binding; sew the binding to the quilt.

4. Add a hanging sleeve, if desired, and label your quilt.

Snowflake Elegance

FINISHED QUILT SIZE: 81" x 81"
FINISHED BLOCK SIZE: 12"

I was not happy with the results when I tried folding fabric and cutting out a snowflake as you do with paper, so I turned instead to uncomplicated shapes. This elongated fleur-de-lis design is simple yet elegant and very easy to work with.

I happened onto these three fabrics in one quilt shop and instantly fell in love. They have a richness I had been looking for.

I used a quick method for appliquéing the snowflakes to the background fabric by using Steam-A-Seam 2. I then used my machine's buttonhole stitch around each appliqué piece. In the process, I discovered that the Steam-A-Seam 2 added just enough stiffness to the appliqué piece and background that I did not need a stabilizer when using the buttonhole stitch. Experiment with your fabric and machine; you may save a step if you don't have stabilizer that you need to remove!

MATERIALS

Yardage is based on 42"-wide fabric.

- 4¼ yards of white-with-gold print for sashing, borders, and binding
- 2⅝ yards of gray print for blocks and cornerstones
- 2¾ yards of white-on-white print for snowflakes and block frames
- 5 yards of fabric for backing
- 85" x 85" piece of batting
- 2½ yards of Steam-A-Seam 2 for appliqué snowflakes
- Stabilizer if needed

CUTTING

All measurements include ¼"-wide seam allowances.

From the white-on-white print, cut:
- 27 strips, 1½" x 42"; crosscut into:
    32 rectangles; 1½" x 12½"
    32 rectangles; 1½" x 14½"
  Use remaining fabric for the snowflake designs.

From the gray print, cut:
- 6 strips, 13" x 42"; crosscut into 16 squares, 13" x 13"
- 2 strips, 3" x 42"; crosscut into 25 squares, 3" x 3"

From the white-with-gold print, cut:
- 20 strips, 3" x 42"; crosscut into 40 rectangles, 3" x 14½"
- 8 strips, 6½" x 42"
- 9 strips, 2½" x 42"

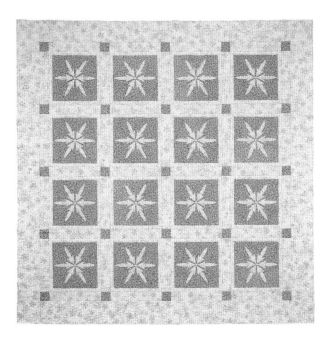

ASSEMBLY

1. Trace 96 of the fleur-de-lis and 16 of the hexagon patterns from page 63 onto the Steam-A-Seam 2.

2. Following the manufacturer's instructions, press the designs to the wrong side of the white-on-white print and cut them out.

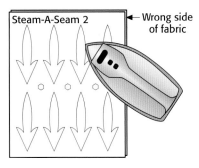

Press.

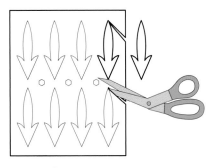

Cut out.

3. To find the center of a 13" gray print square, fold it into quarters and finger-press the folds. When you unfold the square, there will be a vertical and horizontal crease.

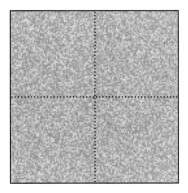

4. Center a hexagon shape over the intersection of the creases on the gray fabric. Align the top and bottom points of the hexagon on the vertical crease.

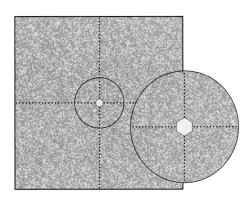

Tip: *Because Steam-A-Seam 2 has a tacky surface, you can reposition the appliqué pieces before you press them down. Simply finger-press them into position and adjust until the appliqués are in the desired place; then press them with your iron for a permanent bond.*

5. Place the six fleur-de-lis sides of the snowflake onto the square using the points of the hexagon as a guide. Adjust the position until you are satisfied with the placement, and press the entire block. Make 16.

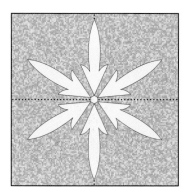

Make 16.

6. You will need to sew the edges of the snowflake down to the background square. I used my machine's buttonhole stitch. Use your favorite machine stitch such as the buttonhole stitch, satin stitch, or blind hem stitch.

TIP: *You can vary the overall look of an appliqué block by changing the color of thread you use to stitch the design. I wanted the snowflake design to be prominent, so I used a thread color that completely blended with the snowflake fabric. The added design element of the decorative stitch is there but in a subtle way. By using a contrasting thread color, you could make a bolder statement and create another facet of the overall design. Experiment!*

7. Trim the blocks to 12½" x 12½" using the hexagon as a center guide.

8. With the block positioned so that 2 segments of the snowflake are vertical, sew the 1½" x 12½" rectangles of white-on-white print to the sides of the blocks. Press the seams toward the rectangles.

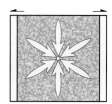

9. Sew the 1½" x 14½" rectangles of white-on-white print to the top and bottom of the blocks. Press the seams toward the rectangles.

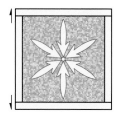

10. Using four blocks per row, create the block rows by sewing the 3" x 14½" rectangles of white-with-gold print to the left side of each block. Join the blocks into rows and sew the remaining rectangles to the right ends of the rows. Press the seams toward the rectangles.

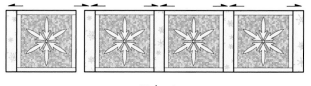

Make 4.

11. Using 4 rectangles per row, create the sashing by sewing a 3" gray square to the left side of each rectangle. Sew the rectangles into rows and then sew the remaining 3" gray squares to the ends of the rows. Press the seams toward the rectangles.

Make 5.

12. Join the block and sashing rows. Press the seams toward the blocks.

BORDERS

1. Sew the 6½" x 42" strips of the white-with-gold print together end to end for the outside border.

2. Measure, cut, and sew the side borders following the instructions in "Quiltmaking Basics" on page 51. Join the side borders to the quilt top. Repeat for the top and bottom borders. Press the seams toward the borders.

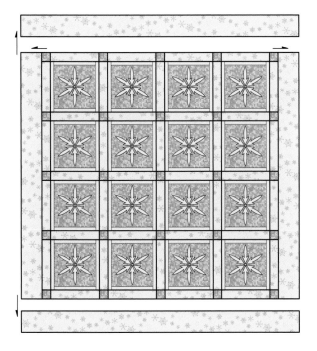

FINISHING

1. Make a quilt sandwich with the quilt top, batting, and backing as shown in "Quiltmaking Basics" on page 51.

2. Quilt as desired. I had this quilt professionally machine quilted by my friend Barb Dau. I love her work and trust her design sense, so we work well together. She brought the snowflake design forward by lightly stippling the gray background fabric.

3. Trim the batting and backing even with the edges of the quilt top. Join the 2½" x 42" white-with-gold print strips end to end for the binding; sew the binding to the quilt.

4. Add a hanging sleeve, if desired, and label your quilt.

TRY THIS! *Use a directional fabric for the snowflakes. Fussy cut each segment of the snowflake to create a uniform look. You may need more fabric if you fussy cut!*

Snowflake Elegance Pillow

Finished Pillow Size: 21½" x 21½"
Finished Block Size: 11½"

Aren't pillows great? They are fun, quick projects that everyone loves to receive as a gift, plus they add that extra sparkle to all decors. Pillows are especially appealing when they coordinate with a quilt. If you are thinking of stitching up the "Snowflake Elegance" quilt on page 32, make one extra block and you'll be able to throw this pillow together as a beautiful companion piece in no time.

MATERIALS

Yardage is based on 42"-wide fabric.

- 1¾ yards of white-with-gold print for outer border and backing
- ⅔ yard of preshrunk muslin (or 23" x 23" square) for pillow back
- ½ yard of gray print for block background
- ½ yard of white-on-white print for the snowflake and inner border
- ⅔ yard of batting (22" x 22")
- 1 sheet 8½" x 11" of Steam-A-Seam 2 or other fusible web for appliqué snowflakes
- ¼ yard of stabilizer, if needed
- 16" pillow form

CUTTING

All measurements include ¼"-wide seam allowances.

From the white-on-white print, cut:
- 4 rectangles, 3" x 24"
  Use remaining fabric for the snowflake designs.

From the gray print, cut:
- 1 square, 13" x 13"

From the white-with-gold print, cut:
- 4 rectangles, 3" x 24"
- 2 rectangles, 22" x 29"

ASSEMBLY

1. To complete the pillow-top center, trace six fleur-de-lis patterns and one hexagon pattern from page 63 onto your fusible web. Follow assembly steps 2–6 from pages 34–35 of the "Snowflake Elegance" instructions.

2. Trim the pillow center to 12" x 12", centering the snowflake motif.

3. Sew one 3" x 24" rectangle of both the white-on-white and white-with-gold prints together along the long edges. Press toward the white-with-gold fabric. Repeat to make 4 strip sets.

Make 4.

4. Fold a strip set in half widthwise and mark with a pin or crease. Center, measure, and mark with a pencil a length of 22¾" along the edge of the white-with-gold fabric. Center, measure, and mark a length of 11¾" along the edge of the white-on-white fabric.

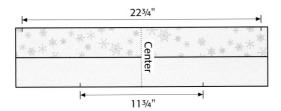

5. Use your ruler to cut the strip set at a 45-degree angle, aligning the edge of the ruler with the pencil marks. Double-check your angle by aligning the 45-degree line of your ruler with the edge of the fabric. Cut the ends of the strip set at opposite angles. Repeat these cuts on the remaining three strip sets.

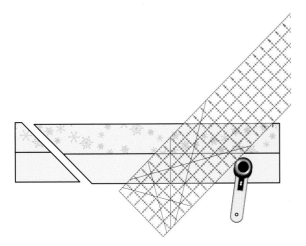

6. To attach the four borders to the center of the pillow top, match the seam intersection points of the block and the border piece and pin. Sew the seam, beginning and ending ¼" from the edge. Press the seams away from the center.

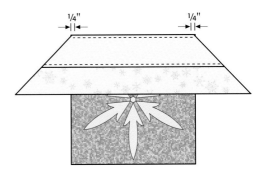

7. With right sides together, fold the corners of the pillow top to form a diagonal fold. With adjacent borders aligned, stitch a ¼" seam from the inside of the borders to the outer edge to create a mitered corner. Repeat for each corner. Press these seams open.

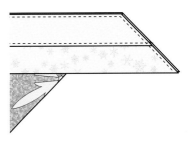

8. Lay the preshrunk muslin square on a flat surface. On top of the muslin, center the batting and smooth out any wrinkles. Center the pillow top over the muslin and batting.

9. Secure the layers together as you would any quilt sandwich. Quilt as desired. I traced around the snowflake with a machine stitch and stitched in the ditch between borders.

Tip: *For small projects—the ones I feel comfortable machine quilting myself—I love using a spray adhesive instead of pinning or basting. My favorite is 505 Spray and Fix; it produces no toxic fumes, so I can use the spray in my home without worry.*

FINISHING

1. Trim the batting and backing even with the edges of the pillow top.

2. With wrong sides together, fold each 22" x 29" rectangle of white-with-gold print in half so that the fold is parallel to the 22" sides. Press, and then stitch 1½" from the folded edge.

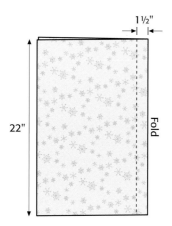

3. Position the two halves of the pillow back over the pillow top. Keep raw edges aligned and overlap the backs. Pin in place and stitch a ¼" seam around the pillow.

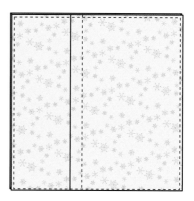

4. Trim the corners and turn the pillow inside out. Press flat and then stitch in the ditch between the inner and outer borders. The pillow form will fit inside this area and the outer border will create a flat flange. You may leave the opening as is or choose your favorite closure such as hook and loop or button. If you choose a button closure, you should reinforce the buttonholes with interfacing and make the buttonholes on one side of the pillow back before you sew the back to the front.

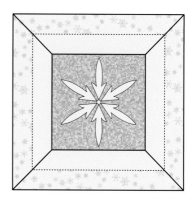

Holiday Cheer!

FINISHED TABLE-RUNNER SIZE: 17½" x 68½"

FINISHED BLOCK SIZE: 12"

I love to decorate for the Christmas season. With warm hearts, my husband, Ed, my daughter, McKenzie, and I feel especially blessed as family and friends gather at our home every year. This simple four-block table runner will be a welcome addition at mealtimes as the candlelight dances across the gold-stitched snowflakes.

Try This! *Stitch up this table runner in a white-and-blue snowflake print and use silver metallic thread for the snowflake quilting to make a great-looking Chanukah table decoration.*

MATERIALS

Yardage is based on 42"-wide fabric.

- ⅞ yard of floral print for blocks and setting triangles
- ½ yard of cream print for blocks and setting triangles
- ¾ yard of green plaid for blocks, setting triangles, and binding
- ⅜ yard of red print for blocks
- 1⅛ yards of fabric for backing
- 21" x 72" piece of batting
- Optional: Gold metallic thread

CUTTING

All measurements include ¼"-wide seam allowances.

From the red print, cut:
- 2 strips, 2½" x 42"; crosscut into 20 squares, 2½" x 2½"
- 1 strip, 3½" x 42"; crosscut into 8 squares, 3½" x 3½"

From the green plaid, cut:
- 1 strip, 2½" x 42"; crosscut into 16 squares, 2½" x 2½"
- 2 strips, 3½" x 42"; crosscut into 14 squares, 3½" x 3½"
- 4 strips, 2½" x 42"

From the floral print, cut:
- 2 strips, 3⅞" x 42"; crosscut into 16 squares, 3⅞" x 3⅞". Cut each square once diagonally to yield 32 half-square triangles.
- 2 strips, 3½" x 42"; crosscut into 12 squares, 3½" x 3½"
- 1 strip, 9¾" x 42"; crosscut into 3 squares, 9¾" x 9¾". Cut each square twice diagonally to yield 12 quarter-square triangles.

From the cream print, cut:
- 1 strip, 7¼" x 42"; crosscut into 4 squares, 7¼" x 7¼". Cut each square twice diagonally to yield 16 quarter-square triangles.
- 1 strip, 3½" x 42"; crosscut into 6 squares, 3½" x 3½"

ASSEMBLY

1. Create nine-patch units with the 2½" squares of red print and green plaid. Press the seams toward the red within each row. Press final seams in either direction.

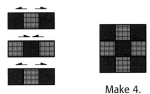

Make 4.

2. Sew a floral-print half-square triangle to each short end of a cream print triangle to create a flying-geese unit as shown. Press the seams toward the corner triangles.

Make 16.

3. Sew a flying-geese unit to two opposite sides of the nine-patch units. Press toward the nine-patch units.

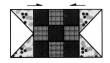

Make 4.

4. Sew the 3½" red print and green plaid squares to opposite ends of the remaining flying-geese units, referring to the illustration below for proper positioning. Press the seams toward the squares.

Make 8.

5. Sew these units to each side of the nine-patch units to complete the blocks. Press the seams away from the nine-patch units.

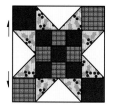

Make 4.

6. Sew the remaining 3½" squares of green plaid, floral print, and cream print together to create four-patch units. Press the seams in each row toward the floral print and press final seams toward the green plaid/floral print row.

Make 6.

7. To create the pieced setting triangles, sew a floral-print quarter-square triangle to two adjacent sides of the four-patch units, referring to the illustration below. Press the seams toward the triangles.

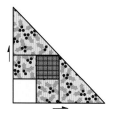

8. Following the diagram, sew a setting triangle between the blocks, forming diagonal rows. There will be only a block at both ends of the runner. Press the seams toward the setting triangles.

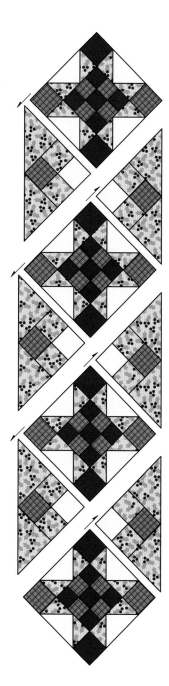

9. Sew the rows together and press the seams in either direction.

FINISHING

1. Make a quilt sandwich with the quilt top, batting, and backing as shown in "Quiltmaking Basics" on page 51.

2. Quilt as desired. I traced seven of the snowflake design on page 63 onto freezer paper and cut them out.

3. Center the snowflake template over the center of each block and the four-patch intersections between the blocks. Press in place.

4. Machine stitch with gold metallic thread around each snowflake.

5. Quilt the rest of the top as desired. I wanted the snowflakes to stand out, so I stitched simple outlines and in the ditch.

6. Trim the batting and backing even with the edges of the quilt top. Join the 2½" x 42" green plaid strips end to end for the binding; sew the binding to the quilt.

7. Add a label to your table runner, and prepare for your holiday feasts!

Designing Your Own Snowflakes

THERE ARE AS many snowflake designs as there are snowflakes. The trick is to make a design that is not too complicated but still striking in appearance. Of course, the fabric you choose will also enhance the design.

As mentioned before, I did not have much success with making a snowflake the old-fashioned way by folding a square piece of fabric and snipping away as you probably did with paper as a child. My efforts just resulted in a lopsided snowflake that looked like it was tortured into existence! I wanted to play but the fabric and scissors were not cooperating.

I then broke down the elements of a snowflake into its individual parts: pretty spires that emanate from the center, and crystallized shoots like delicate fingers that reach out from the spires. Some snowflakes are chubby with the crystals engulfing the spires to make a beautiful hexagon shape, while some are skinny with just the spires stretching out from the center.

So, taking the basic shape and elements of a snowflake, I started playing with simple, everyday shapes to form a snowflake of my own design. Squares, circles, diamonds, and stars are what I have to show you here, but don't stop there! Any simple shape will do—add hearts in different sizes, or rectangles or triangles, and mix the shapes together in one snowflake. Your imagination is your only limitation. If you need inspiration, do what I do. I use kids! I had the most wonderful experience one very rainy and nasty Sunday a

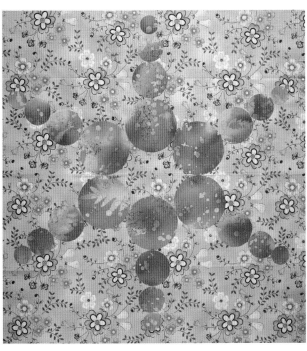

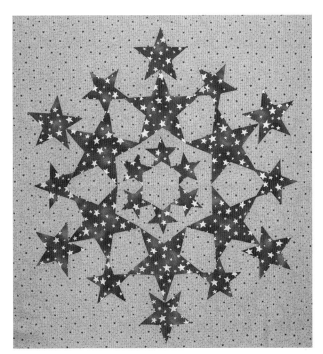

while back. I was taking care of my best friend's kids, Daniel and Madeleine, for the weekend. They were bored, couldn't go outside because of the weather, and I had work to do so I wasn't free to entertain them just then. I pulled out my colored pencils and several photocopies of line-drawn quilt blocks and asked them to color them. I was overwhelmed by the results! Not only were these very active kids pleasantly occupied for hours, but they also began to challenge themselves creatively. Madeleine would close her eyes and randomly pick out three colored pencils and *have* to use them and Daniel, the younger of the two, discovered that he could use more than one color for the blocks. They were very pleased with themselves, as was I. It was a fabulous afternoon! I selfishly asked to keep their work and they said yes. I hope to incorporate their work in a book someday.

So if you would like to have as much fun, draw simple shapes or use the ones I have provided for you (page 49) on the Steam-A-Seam 2, fuse them to fabric, cut them out, and cut some squares of fabric for backgrounds. Grab some kids aged 6–12, whether they are yours or not, show them what to do, and let them play. You will be amazed at what they come up with! Fuse the shapes to the background fabric; add a border or two, quilt it up, and you have a wonderful quilt to give to the creator. Their designs will give you all sorts of ideas and inspiration!

You will need about ¼ yard of fabric for the designs and one fat quarter of fabric for the back-ground if you are using shapes in the sizes that are provided in this book.

Choose the segment designs that are on page 49 or design your own. I have used the segment shapes in graduating sizes to add a little extra dimension to my blocks. You will need six of each size. Trace and cut the segments out of the snowflake design fabric using Steam-A-Seam 2 as described on page 20 to fuse the segments to the background fabric, or use your favorite appliqué method.

When designing your snowflake block, use the placement template on page 50. Trace it onto a large piece of paper and extend each line so that as you are placing the segments of a snowflake onto the background fabric you can use the placement template as a guide. Use the guide with a light box and place it under the background fabric. You may also lightly trace the guide onto the background fabric, making sure the snowflake segments completely cover the pencil line.

The space between each spire is 60 degrees. After designing one or two snowflakes, you will be able to establish the center of the background fabric as described in step 3 on page 34, work your first straight up-and-down spires, and fill in the rest by eye. It really is that simple. The size of the segments will help you place them on the background fabric because they all have to fit evenly around the center point. One and one-half segments must fit evenly within a 90-degree area as shown opposite.

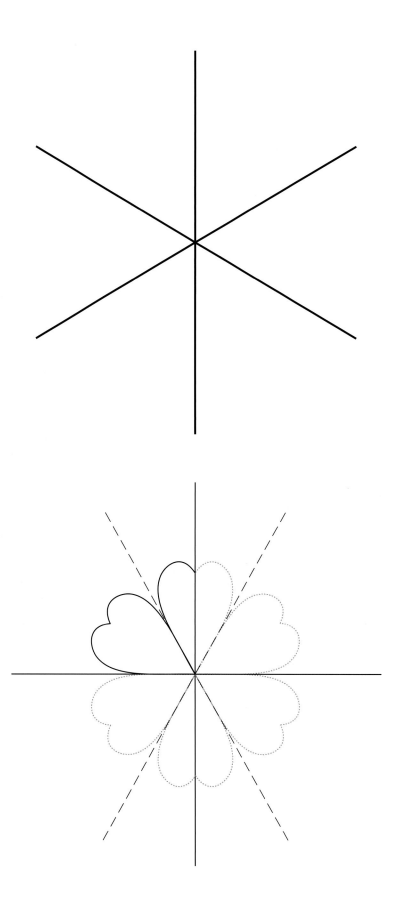

Quiltmaking Basics

This section provides you with all the necessary elements for successfully completing your project, from choosing fabric and assembling essential tools for quiltmaking, to making embroidery stitches, squaring up blocks, adding borders, making a quilt sandwich, quilting, binding, and finally, adding a hanging sleeve and label.

FABRICS

Fabric, glorious fabric. I just love fabric! It is a real joy to step into a quilt shop or fabric store, my home away from home. I find it amazing what today's fabric manufacturers are able to do with color and design; it is so much different from when I started sewing in the late 1960s.

For quilting I use 100% cotton fabric. I try to buy the best-quality cotton my budget can afford but that doesn't mean I don't look through the bargain bins—some of my favorite fabrics are the inexpensive finds. Many people think cheaper fabrics won't last as long and I can't tell you whether that is true or not. It is my philosophy that it's OK if my quilts don't last generations. I want my quilts to be loved, used, and used up. I have the privilege of being able to make many, many quilts; some will survive and some won't. This thinking is very freeing for me, giving me the opportunity to use what I want, even when it came from the bargain bin.

Yardage requirements are provided for all projects in this book and are based on 42" of usable fabric, with the exception of the "Snow in the Mountains" project on page 27. For that project I used flannel fabric that I preshrunk, so I based the requirements on 40" of usable fabric.

To prewash or not prewash, that is the question, and one that sparks constant debate among quilters. I was teaching a class a while back and got a skeptical look from a couple of my students while the others were nodding in agreement. Can you guess what I was discussing? You guessed it, I don't prewash my fabric unless I am using muslin or flannel. Instead of going into the reasons why I don't prewash, I would like to leave it up to the individual. Whatever you are comfortable doing, do it!

SUPPLIES

Sewing machine: You will need a sewing machine that has a good straight stitch. You'll also need a walking foot or darning foot if you are going to machine quilt. Take a moment to clean and oil your machine. Try to get into the practice of cleaning your machine before every project. Cotton is a great fiber but it does create lint under the feed dogs that can interfere with the smooth running of your sewing machine.

Rotary-cutting tools: You will need a rotary cutter, a cutting mat, and a clear acrylic ruler. The 6" x 24" size works well for cutting long strips and squares. You should also have a large square ruler for squaring up quilt blocks.

Thread: Use a good-quality, all-purpose cotton or cotton-covered polyester thread. Choose a neutral color such as gray so that the thread won't show through when you're piecing light and dark fabrics together.

Additional tools: You will need hand and machine sewing needles, pins, and fabric scissors for cutting threads. Don't forget the seam ripper, the smaller the better so the point can slide through the stitches easily. You'll also need an iron and ironing board, as well as tracing tools such as pencil and paper for the appliqué projects.

Fusible web: I use Steam-A-Seam 2 by the Warm Company. It holds very well and adds just enough body to the fabric so that I generally don't have to use a stabilizer when I blanket stitch the edges by machine. It also does not make the fabric too stiff.

ROTARY CUTTING

All the projects in this book are designed for quick-and-easy rotary cutting, except for the pieces involved in fusible appliqué. All measurements include standard ¼"-wide seam allowances, but the templates do not. Here is a quick lesson on rotary cutting.

1. Fold the fabric and match the selvages, aligning the crosswise and lengthwise grains as much as possible. Place the folded edge closest to you on the cutting mat. Align a square ruler along the folded edge of the fabric. Then place a long, straight ruler to the left of the square ruler, just covering the uneven raw edges of the left side of the fabric.

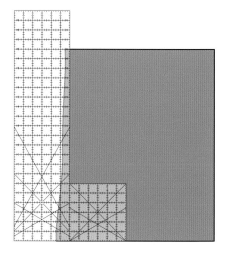

   Remove the square ruler and cut along the right edge of the long ruler, rolling the rotary cutter away from you. Discard this strip. (Reverse this procedure if you are left-handed.)

2. To cut strips, align the required measurement on the ruler with the newly cut edge of the fabric. For example, to cut a 3"-wide strip, place the 3" ruler mark on the edge of the fabric.

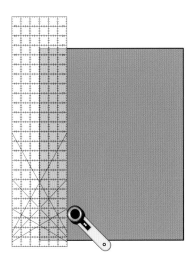

3. To cut squares, cut strips in the required widths. Trim away the selvage ends of the strips. Align the required measurement on the ruler with the left edge of the strip and cut a square. Continue cutting squares until you have the number you need.

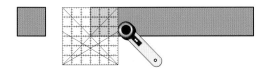

MACHINE PIECING

The most important thing to remember about machine piecing is to maintain a consistent ¼"-wide seam allowance. Some sewing machines have a special ¼" foot that measures exactly ¼" from the center needle position to the edge of the foot. This feature allows you to use the edge of the presser foot to guide the fabric for a perfect ¼"-wide seam allowance.

If your machine doesn't have such a foot, create a seam guide by placing the edge of a piece of tape ¼" away from the needle.

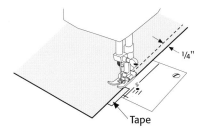

CHAIN-PIECING

Chain-piecing is an efficient, timesaving system.

1. Sew the first pair of pieces from cut edge to cut edge, using about 12 stitches per inch. At the end of the seam, stop sewing, but do not cut the thread.

2. Feed the next pair of pieces under the presser foot, as close as possible to the first pair. Continue feeding pieces through the machine without cutting the thread in between. There is no need to backstitch, since each seam will be crossed and held by another seam.

3. When all pieces have been sewn, remove the chain from the machine and clip the threads between pieces.

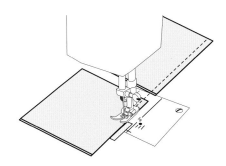

EASING

If two pieces that will be sewn together are slightly different in size (by less than ⅛"), pin the places where the two pieces should match. Next pin the middle, if necessary, to distribute the excess fabric evenly. Sew the seam with the longer piece on the bottom, next to the feed dogs. The feed dogs will help ease the two pieces together.

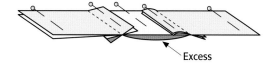

Excess

PRESSING

Pressing is important for several reasons: it helps sink the thread into the fabric, it sets the seam, and with a properly pressed seam you will be able to tell if the pieces are accurate. Be aware that pressing and ironing are two different things. While pressing, use an up-and-down motion rather than sliding the iron back and forth over the block. This will help prevent distortion. I use both a dry iron and steam, depending on how the fabric reacts to dry or steam heat. Make sure the iron face is clean.

The traditional rule in quiltmaking is to press the seams to one side, toward the darker color wherever possible. Press the seam flat from the wrong side first, and then press the seam in the desired direction from the right side. Be particularly careful when pressing bias seams or edges.

When joining two seamed units, plan ahead and press the seam allowances in the opposite directions as shown. This reduces bulk and makes it easier to match seam lines. Where two seams meet, the seam allowances will butt against each other, making it easier to join units with perfectly matched seam intersections. All the projects include illustrations with pressing arrows that

show which way to press the seams so that they'll butt against each other properly.

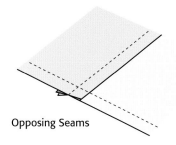

Opposing Seams

APPLIQUÉ

I want my projects to be quick and easy, so I use the double-backed fusible web for the appliqué projects in this book. Please use your favorite appliqué method. The templates that are given for each of the projects are without seam allowance.

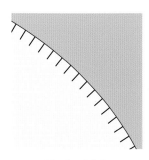

Blanket Stitch

If you do use fusible web, follow the directions on the package. For any project that will be handled a lot, you will need to anchor the edges of the appliqué so they won't lift up with lots of washing. I use my sewing machine's programmed blanket stitch to secure the appliqué to the background fabric. You can also use a blind hem stitch or satin stitch by machine. You may also

Blind Hem Stitch

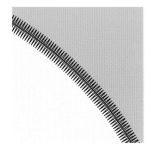

Satin Stitch

use a running stitch by hand for a primitive look but remember that the raw edges may fray and lift when washed or handled repeatedly.

EMBROIDERY STITCH

The stitch I would recommend for the "Snowflakes in Redwork" project on page 22 is the stem stitch. Although you may choose to back-stitch as a slightly easier alternative, the stem stitch adds a little extra bulk and dimension to the design and it helps to hide the small wobbles in my stitching.

Use two strands of floss about 18" long, separate the strands, and put them back together before you start stitching.

Don't knot the end of the floss before you begin. As you begin, bring up the needle and thread from the back, anchor the end with your finger, and simply stitch over the ends a couple of times. Clip the end of the floss if there is too much left.

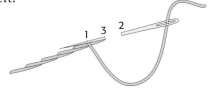

Stem Stitch

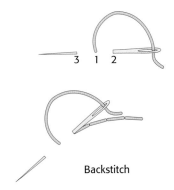

Backstitch

SQUARING UP BLOCKS

There are a couple of projects that require you to cut background squares that are larger than actually needed. You will be manipulating the squares with embroidery or appliqué and they may become distorted. Squaring up the blocks will give you a perfectly centered motif and set the accuracy for the rest of the project.

Use a large square ruler to measure and trim the blocks. Make sure you trim all sides of the block or your block will be lopsided.

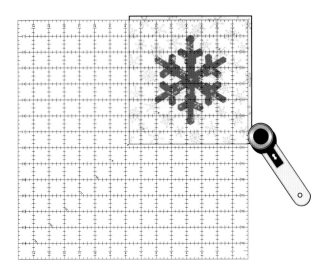

If your blocks are not the size required for the project, simply adjust all of the other components, such as sashings and borders, accordingly.

BORDERS

For best results, do not cut border strips and sew them directly to the quilt sides without measuring first unless the quilt is small, such as a wall hanging. The edges of a quilt often measure slightly longer than the distance through the quilt center, due to stretching during construction. Instead, measure the quilt top through the center in both directions to determine how long to cut the border strips. This step ensures that the finished quilt will be as straight and as "square" as possible, without wavy edges.

Plain border strips are cut along the crosswise grain and seamed where extra length is needed.

Straight-Cut Borders

All of the borders in this book are either straight-cut borders or have details of one kind or another. For the detailed borders, please follow the directions given in the project instructions.

1. Measure the length of the quilt top through the center. Cut border strips to that measurement, piecing as necessary; mark the center of the quilt edges and the border strips. Pin the borders to the sides of the quilt top, matching the center marks and ends and easing as necessary. Sew the border strips in place. Press the seams toward the border.

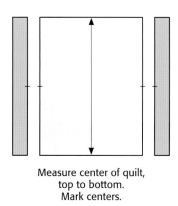

Measure center of quilt,
top to bottom.
Mark centers.

2. Measure the width of the quilt top through the center, including the side borders just added. Cut border strips to that measurement, piecing as necessary; mark the center of the quilt edges and the border strips. Pin the borders to the top and bottom edges of the quilt top, matching the center marks and ends and easing as necessary; stitch. Press the seams toward the border.

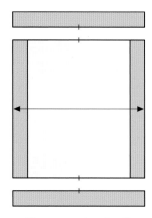

Measure center of quilt,
side to side, including borders.
Mark centers.

Mitered Borders

The only project in this book that includes mitered borders is the "Snowflake Elegance Pillow" on page 37, and the instructions to create a mitered border are included.

MARKING QUILTING LINES

Whether or not to mark the quilting designs depends upon the type of quilting you will be doing. Marking is not necessary if you plan to quilt in the ditch or outline quilt a uniform distance from seam lines. For more complex quilting designs, mark the quilt top before the quilt is layered with batting and backing.

Choose a marking tool that will be visible on your fabric and test it on fabric scraps to be sure the marks can be removed easily. Masking tape can be used to mark straight quilting. Tape only small sections at a time and remove the tape when you stop at the end of the day; otherwise, the sticky residue may be difficult to remove from the fabric.

LAYERING THE QUILT

The quilt "sandwich" consists of backing, batting, and the quilt top. Cut the quilt backing several inches larger than the quilt top all the way around. For large quilts, it is usually necessary to sew two or three lengths of fabric together to make a backing of the required size. Trim away the selvages before piecing the lengths together. Press the seams open to make quilting easier.

1
fabric
width

Two lengths of fabric
seamed in the center

Partial fabric width

Batting comes packaged in standard bed sizes, or it can be purchased by the yard. Several weights or thicknesses are available. Thick battings are fine for tied quilts and comforters; a thinner batting is better, however, if you intend to quilt by hand or machine.

To put it all together:

1. Spread the backing, wrong side up, on a flat, clean surface. Anchor it with pins or masking tape. Be careful not to stretch the backing out of shape.

2. Spread the batting over the backing, smoothing out any wrinkles.

3. Place the pressed quilt top, right side up, on top of the batting. Smooth out any wrinkles and make sure the quilt-top edges are parallel to the edges of the backing.

4. Starting in the center, baste with needle and thread and work diagonally to each corner. Continue basting in a grid of horizontal and vertical lines 6" to 8" apart. Finish by basting around the edges.

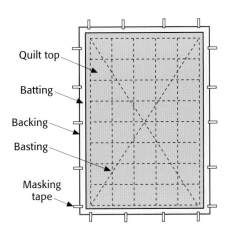

Quilt top
Batting
Backing
Basting
Masking tape

NOTE: *For machine quilting, you may baste the layers with rustproof safety pins. Place pins about 6" to 8" apart, away from the area you intend to quilt.*

TIP: *For small projects that I plan to machine quilt, I have also used a spray adhesive to anchor the layers together with a lot of success! There are several brands available on the market but my favorite is 505 Spray and Fix; it doesn't produce toxic fumes and works very well.*

MACHINE QUILTING

Machine quilting is suitable for all types of quilts, from crib to full-size bed quilts. With machine quilting, you can quickly complete quilts and start that next project that is calling out to you.

Marking is only necessary if you need to follow a grid or a complex pattern. It is not necessary if you plan to quilt in the ditch, outline quilt a uniform distance from seam lines, or free-motion quilt in a random pattern over the quilt surface or in selected areas.

1. For straight-line quilting, it is extremely helpful to have a walking foot to help feed the quilt layers through the machine without shifting or puckering. Some machines have a built-in walking foot; other machines require a separate attachment.

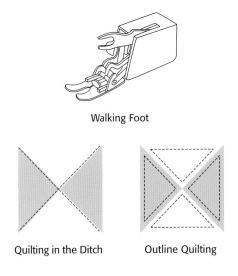

Walking Foot

Quilting in the Ditch

Outline Quilting

2. For free-motion quilting, you need a darning foot and the ability to drop the feed dogs on your machine. With free-motion quilting, you do not turn the fabric under the needle but instead guide the fabric in the direction of the design. Use free-motion quilting to outline-quilt a fabric motif or to create stippling or other curved designs.

Darning Foot

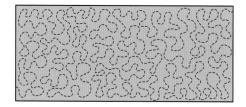

Free-Motion Quilting

ADDING A SLEEVE

If you plan to display your finished quilt on the wall, be sure to add a hanging sleeve to hold the rod.

1. Using leftover fabric from the front or a piece of muslin, cut a strip 6" to 8" wide and 1" shorter than the width of the quilt at the top edge. Fold the ends under ½", then ½" again; stitch.

2. Fold the fabric in half lengthwise, wrong sides together, and baste the raw edges to the top edge of the quilt back. The top edge of the sleeve will be secured when the binding is sewn on the quilt.

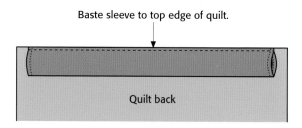

Baste sleeve to top edge of quilt.

Quilt back

3. Finish the sleeve after the binding has been attached by blindstitching the bottom of the sleeve in place. Push the bottom edge of the sleeve up just a bit to provide a little give so that the hanging rod does not put strain on the quilt itself.

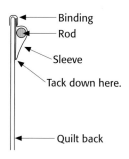

Binding
Rod
Sleeve
Tack down here.

Quilt back

BINDING

Bindings can be made from straight-grain or bias-grain strips of fabric. All of the quilts in this book call for a French double-fold binding, cutting strips at 2½" wide.

To cut straight-grain binding strips, cut 2½"-wide strips across the width of the fabric. You will need enough strips to go around the perimeter of the quilt plus 10" for seams and the corners in a mitered fold.

1. Sew strips, right sides together, to make one long piece of binding. Press the seams open. Join strips at right angles and stitch across the corner as shown. Trim excess fabric and press the seams open.

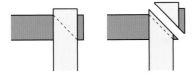

Joining Straight-Cut Strips

2. Trim one end of the strip at a 45-degree angle, turn under ¼", and press. Trimming the end at an angle distributes the bulk so you won't have a lump where the two ends of the binding meet. Fold the strip in half lengthwise, wrong sides together, and press.

Fold line

3. Trim the batting and backing even with the quilt top. If you plan to add a sleeve, do so now before attaching the binding (see page 58).

4. Starting on one side of the quilt and using a ¼"-wide seam allowance, stitch the binding to the quilt, keeping the raw edges even with the edge of the quilt top. End the stitching ¼" from the corner of the quilt and backstitch. Clip the thread.

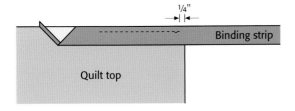

¼"

Binding strip

Quilt top

5. Turn the quilt so that you'll be stitching down the next side. Fold the binding up, away from the quilt, then back down onto itself, parallel with the edge of the quilt top. Begin stitching at the edge, backstitching to secure. Repeat on the remaining edges and corners of the quilt.

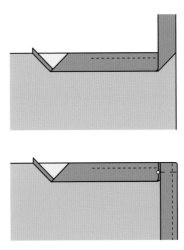

6. When you reach the beginning of the binding, overlap the beginning stitches by 1" and cut away any excess binding, trimming the end at a 45-degree angle. Tuck the end of the binding into the fold and finish the seam.

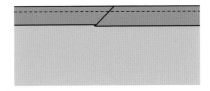

7. Fold the binding over the raw edges of the quilt to the back, with the folded edge covering the row of machine stitching, and blindstitch in place. A miter will form at each corner. Blindstitch the mitered corners.

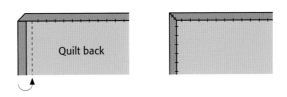

Quilt back

SIGNING YOUR QUILT

Be sure to sign and date your quilt. Future generations will be interested to know more than just who made it and when. Labels can be as elaborate or as simple as you desire. The information can be handwritten, typed, or embroidered. Be sure to include the name of the quilt, your name, your city and state, the date, the name of the recipient if it is a gift, and any other interesting or important information about the quilt.

Patterns

Snowflakes in Redwork

Snowflakes in Redwork

**Puttin' on the Ritz
Holiday Cheer!**

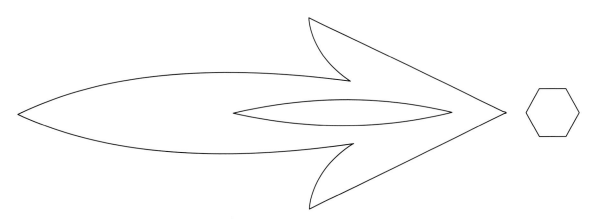

Snowflake Elegance

About the Author

TERRY MARTIN began her love affair with sewing as a little girl, when her "Gammy" put a needle and thread in her hands. At first, her passion was for garment making, since that was the only way for a chubby girl to have fashionable clothes at the time! Terry moved on to needlepoint and counted cross-stitch—but once she discovered quilting, she went over the creative edge and never came back.

Terry lives in Snohomish, Washington, with husband Ed, daughter McKenzie, and four black cats. She also lectures, teaches, and is the author of two previous books. Terry is the editorial assistant and author liaison at Martingale & Company.